12 THEORIES OF ALIEN LIFE: EXPLORING THE POSSIBILITIES BEYOND EARTH

By Rayan Bale

Your Gateway to the Stars

Copyright © 2024 by Rayan Bale

All rights reserved. No part of this publication may be reproduced, distributed, or transmitted in any form or by any means, including photocopying, recording, or other electronic or mechanical methods, without the prior written permission of the publisher, except in the case of brief quotations embodied in critical reviews and certain other noncommercial uses permitted by copyright law.

TABLE OF CONTENTS

Introduction　　　　　　　　　　　　　　　3

Theory 1: The Drake Equation:　　　　　　10
　　Estimating our cosmic neighbors.

Theory 2: The Fermi Paradox:　　　　　　　18
　　Where is everybody?

Theory 3: The Rare Earth Hypothesis:　　　27
　　Complex life is a rare gem.

Theory 4: The Aestivation Hypothesis:　　　34
　　Hibernating until the universe cools.

Theory 5: The Zoo Hypothesis:　　　　　　 42
　　Observed but untouched.

Theory 6: Panspermia:　　　　　　　　　　49
　　Seeds of life scattered across the stars.

Theory 7: Gaian Bottleneck Hypothesis:　　 57
　　Surviving the cosmic struggle.

Theory 8: Technosignatures: 65
    Finding alien technology, not just biology.

Theory 9: Trapped in Deep Oceans: 76
    Life hidden under icy shells.

Theory 10: Hydrogen-Breathing Life Forms 86
    Living without oxygen.

Theory 11: Black Hole Habitats: 95
    Thriving near cosmic giants.

Theory 12: The Transient Hypothesis: 101
    Briefly existing civilizations.

Conclusion 109

Glossary of Terms 114

References 118

INTRODUCTION

The Quest for Extraterrestrial Life: A Journey Beyond Earth:

From the dawn of civilization, humans have gazed at the stars and wondered about the existence of life beyond our blue planet. The night sky, with its countless stars and mysterious celestial bodies, has inspired myths, legends, and scientific inquiry alike. As our understanding of the cosmos has expanded, so too has our curiosity about what—or who—might be out there.

In the vast expanse of the universe, with its billions of galaxies, each containing billions of stars and potentially even more planets, the probability of life existing elsewhere seems not just possible but probable. Yet, despite our technological advancements and dedicated efforts, the definitive evidence of extraterrestrial life remains elusive. This book delves into the fascinating and complex theories that scientists and thinkers have proposed to explain this great mystery.

Why Search for Extraterrestrial Life?

The search for extraterrestrial life is not just about answering a fundamental question of our place in the universe; it's also about understanding life itself. By exploring the possibilities of life beyond Earth, we gain insights into the origins, adaptability, and resilience of life in the most extreme conditions. This quest drives innovation in science and technology, pushing the boundaries of what we know and what we can achieve.

The Structure of This Book:

In "12 Theories of Alien Life: Exploring the Possibilities Beyond Earth," we will embark on a journey through twelve compelling theories that seek to explain the potential existence and nature of extraterrestrial life. Each chapter will delve into a specific theory, examining the scientific principles, evidence, and speculations that underpin it. From the statistical probabilities of the Drake Equation to the philosophical ponderings of the Fermi Paradox, and from the intriguing possibilities of subsurface oceans on distant moons to the speculative nature of hydrogen-breathing organisms, this book covers a broad spectrum of scientific thought.

Here is a brief overview of the twelve theories we will explore:

1-The Drake Equation: Formulated by Dr. Frank Drake, this equation estimates the number of active, communicative extraterrestrial civilizations in the Milky Way galaxy.

2-The Fermi Paradox: Named after physicist Enrico Fermi, this paradox highlights the apparent contradiction between the high probability of extraterrestrial civilizations and the lack of evidence for, or contact with, such civilizations.

3-The Rare Earth Hypothesis: This theory posits that while microbial life might be common in the universe, complex life forms are extremely rare and Earth-like planets are unique.

4-The Aestivation Hypothesis: Suggests that advanced alien civilizations might be in a state of hibernation until environmental conditions in the universe become more favorable for their activities.

5-The Zoo Hypothesis: Proposes that extraterrestrial civilizations intentionally avoid contact with us to allow for natural evolution and sociocultural development, similar to zookeepers observing animals.

6-Panspermia: This theory suggests that life exists throughout the universe and is distributed by meteoroids, asteroids, comets, and planetoids.

7-Gaian Bottleneck Hypothesis: Argues that life needs specific environmental conditions to develop and sustain itself, which are incredibly rare. Most life forms fail to survive long enough to evolve into complex beings.

8-Technosignatures: The search for alien technology or artifacts, such as Dyson spheres or advanced communication signals, which could indicate the presence of intelligent life.

9-Trapped in Deep Oceans: The idea that life could exist in subsurface oceans on moons like Europa and Enceladus, protected from harsh surface conditions by thick ice layers.

10-Hydrogen-Breathing Life Forms: Suggests that alien life forms might not require oxygen but could thrive on other gases like hydrogen, which are more abundant in the universe.

11-Black Hole Habitats: The concept that life could exist in the extreme conditions near black holes, using the energy from the accretion disks or relativistic jets as a source of warmth and sustenance.

12-The Transient Hypothesis: Proposes that we have not found extraterrestrial life because it is transient in nature, either due to rapid technological advancement leading to self-destruction or other existential risks.

The Importance of Open Inquiry:

As we explore these theories, it is crucial to maintain an open mind. The search for extraterrestrial life is inherently interdisciplinary, drawing from astronomy, biology, physics, chemistry, and even sociology and philosophy. Each theory offers a unique lens through which we can view the cosmos and our place within it. Some theories may seem more plausible than others, and some may challenge our preconceived notions, but all contribute to the rich tapestry of our understanding.

A Glimpse into the Unknown:

Our journey begins with the Drake Equation, a mathematical framework that estimates the number of active, communicative extraterrestrial civilizations in the Milky Way galaxy. From there, we will traverse a landscape of scientific inquiry and imaginative speculation, exploring the myriad ways life might arise and thrive in the universe. Whether life is common or rare, microbial or intelligent, visible or hidden beneath icy crusts, each theory adds a piece to the cosmic puzzle.

Join us as we embark on this exploration of the unknown, guided by the light of science and the boundless curiosity that defines our species. Together, we will ponder the profound questions and thrilling possibilities that lie beyond the stars, seeking to uncover the secrets of life in the vast universe.

THEORY I:
THE DRAKE EQUATION:
ESTIMATING OUR COSMIC NEIGHBORS.

Introduction to the Drake Equation:

Formulated by Dr. Frank Drake in 1961, the Drake Equation was one of the first scientific attempts to estimate the number of active, communicative extraterrestrial civilizations in the Milky Way galaxy. This groundbreaking formula has since become a foundational tool in the field of astrobiology and the search for extraterrestrial intelligence (SETI). The equation has sparked countless debates, research initiatives, and a broader understanding of our place in the universe. Its elegance lies in its ability to break down the complex question of extraterrestrial life into manageable scientific components, each of which can be studied and refined as our knowledge grows.

Components of the Drake Equation:

The Drake Equation is expressed as:

$$N = R_{*} \times f_p \times n_e \times f_l \times f_i \times f_c \times L$$

Where:

- **N:** The number of civilizations in our galaxy with which humans could communicate.
- **R*:** The average rate of star formation per year in our galaxy.
- **f_p:** The fraction of those stars that have planetary systems.
- **n_e:** The average number of planets that could potentially support life per star that has planets.
- **f_l:** The fraction of planets that could support life where life actually appears.
- **f_i:** The fraction of planets with life where intelligent life evolves.
- **f_c:** The fraction of civilizations that develop a technology that releases detectable signs of their existence into space.
- **L:** The length of time such civilizations can communicate.

Each term in the equation represents a critical step in the development of communicative civilizations. By breaking down the question of extraterrestrial life into these components, the Drake Equation allows scientists to tackle each piece of the puzzle individually, refining their estimates as new data becomes available. This modular approach is part of the equation's enduring appeal and utility.

Historical Context and Dr. Frank Drake's Contributions:

Dr. Frank Drake first introduced this equation during a meeting at the National Radio Astronomy Observatory in Green Bank, West Virginia. The purpose was to quantify the variables involved in finding extraterrestrial life and to stimulate scientific discussion. At the time, the values for these parameters were largely unknown, leading to wide-ranging estimates for the number of communicative civilizations. This uncertainty was not a flaw but rather a feature of the equation, highlighting the gaps in our knowledge and guiding future research efforts.

Drake's equation has since inspired countless research efforts and has been a driving force behind the SETI project, which uses radio telescopes to listen for signals from intelligent extraterrestrial sources. Drake's work laid the foundation for modern astrobiology, influencing how scientists think about the search for life beyond Earth. His contributions extend beyond the equation itself, encompassing a broader philosophy of scientific exploration and the pursuit of knowledge.

Modern Applications and Revisions:

Since its formulation, the Drake Equation has been revised and expanded as our understanding of the cosmos has grown. The discovery of thousands of exoplanets by missions such as NASA's Kepler and TESS (Transiting Exoplanet Survey Satellite) has provided new data for some of the equation's parameters. For example, we now know that many stars have planetary systems, and a significant fraction of these planets are in the habitable zone where liquid water could exist.

Recent estimates suggest that there could be billions of Earth-like planets in our galaxy alone. However, the values for the latter parameters, especially those related to the development and longevity of intelligent life, remain speculative. Despite this, the Drake Equation continues to be a valuable framework for guiding scientific inquiry and refining our search strategies. It serves as a reminder of both the progress we have made and the mysteries that still elude us.

Example Calculation Using the Drake Equation:

To illustrate how the Drake Equation works, let's use some hypothetical values for each parameter:

- **R*:** 1 star per year
- **f_p:** 0.5 (50% of stars have planetary systems)
- **n_e:** 2 (each planetary system has 2 planets capable of supporting life)
- **f_l:** 0.33 (one-third of those planets develop life)
- **f_i:** 0.01 (1% of planets with life develop intelligent life)
- **f_c:** 0.1 (10% of civilizations develop detectable technology)
- **L:** 10,000 years (civilizations can communicate for 10,000 years)

Plugging these values into the equation, we get:

$$N = 1 \times 0.5 \times 2 \times 0.33 \times 0.01 \times 0.1 \times 10{,}000 = 3.3$$

This calculation suggests that there could be approximately 3.3 civilizations in the Milky Way galaxy with which we might be able to communicate. This is a simplified example, but it demonstrates how the equation can be used to make educated guesses about the prevalence of extraterrestrial civilizations based on current astronomical and biological knowledge.

Case Studies: The Use of the Drake Equation in Recent Research:

One notable application of the Drake Equation is in the assessment of the potential for finding extraterrestrial life with new technologies. For example, the SETI Institute uses the equation to prioritize target stars for their listening efforts. Advances in radio astronomy and the development of new observatories, such as the Square Kilometre Array (SKA), are expected to enhance our ability to detect faint signals from distant civilizations.

In another instance, the Breakthrough Listen initiative, a comprehensive search for intelligent life, has utilized the Drake Equation to frame its goals and expectations. This project, funded by philanthropist Yuri Milner, aims to scan the nearest million stars and 100 galaxies for signs of technology, using some of the world's most powerful telescopes. These case studies demonstrate the equation's ongoing relevance and its capacity to adapt to new scientific tools and discoveries.

Philosophical Implications:

Beyond its scientific utility, the Drake Equation has profound philosophical implications. It challenges us to think about our place in the universe and the possibility that we are not alone. The equation also raises questions about the longevity of civilizations, including our own, and what factors might influence the survival and communication capabilities of intelligent life. These reflections go beyond empirical data, touching on the broader human experience and our quest for meaning in a vast and mysterious cosmos.

Conclusion:

The Drake Equation remains a cornerstone in the search for extraterrestrial life. While many of its parameters are still uncertain, it provides a structured way to consider the vast array of factors involved in finding intelligent life beyond Earth. As our technological capabilities advance and our knowledge of the universe expands, we continue to refine the equation and our search strategies, driven by the enduring hope of answering one of humanity's most profound questions: Are we alone in the universe?

THEORY 2:
THE FERMI PARADOX

WHERE IS EVERYBODY?

Introduction to the Fermi Paradox:

The Fermi Paradox, named after the Italian-American physicist Enrico Fermi, highlights the apparent contradiction between the high probability of extraterrestrial civilizations' existence and the lack of evidence for, or contact with, such civilizations. Fermi famously posed the question, "Where is everybody?" during a casual lunchtime conversation with colleagues at Los Alamos National Laboratory in 1950. This paradox has since become a central topic in the search for extraterrestrial intelligence (SETI) and continues to provoke thought and debate among scientists and philosophers. The paradox not only addresses the silence of the cosmos but also challenges our assumptions about life, intelligence, and the future of civilizations.

The Statistical Likelihood of Extraterrestrial Life:

Given the vast number of stars in the Milky Way galaxy—estimated to be around 100 billion—and the increasing number of exoplanets discovered within their habitable zones, the statistical likelihood of extraterrestrial life seems high. The principles of the Copernican revolution suggest that Earth is not unique; rather, it is one of many planets that could harbor life. If even a tiny fraction of these planets develop intelligent civilizations, the number of such civilizations should be substantial. The Fermi Paradox arises from this high probability contrasted with the lack of observable evidence. This discrepancy becomes even more perplexing when considering the advanced age of the galaxy, which has had ample time to foster numerous civilizations.

Possible Resolutions to the Fermi Paradox:

Numerous hypotheses have been proposed to explain the Fermi Paradox. Here are some of the most compelling:

Rare Earth Hypothesis: This hypothesis suggests that while microbial life might be common in the universe, complex life forms are extremely rare and Earth-like planets are unique due to a combination of rare geological and environmental conditions. The delicate balance of factors such as tectonic activity, a large moon, a magnetic field, and a stable climate might be exceedingly uncommon, making Earth a rare oasis in a vast cosmic desert. (This theory will be discussed in more detail in Theory 3.)

Great Filter: This concept posits that there is a stage in the evolution of life that is extremely difficult to surpass. This "Great Filter" could be behind us, meaning that the emergence of intelligent life is exceptionally rare, or ahead of us, suggesting that advanced civilizations inevitably self-destruct. The Great Filter theory raises profound questions about the future of humanity and our ability to navigate existential threats such as nuclear war, environmental collapse, or unchecked artificial intelligence.

Percolation Theory: This theory, borrowed from the study of networks, suggests that civilizations may only expand so far before they encounter diminishing returns or existential risks, leading to a fractal-like pattern of isolated civilizations that never meet. This model implies that the spread of civilizations might be limited by natural barriers, resource limitations, or strategic decisions to avoid conflict.

The Zoo Hypothesis: This hypothesis proposes that extraterrestrial civilizations intentionally avoid contact with us to allow for natural evolution and sociocultural development, similar to zookeepers observing animals without interference. Advanced civilizations might consider humanity not yet ready for direct contact, opting instead to observe us from a distance, possibly through sophisticated monitoring technologies that remain undetectable by our current means. (This theory will be discussed in more detail in Theory 5.)

Self-Imposed Isolation: Advanced civilizations might choose to isolate themselves to preserve their cultures, avoid potential threats, or focus on virtual realities rather than physical expansion. This self-imposed isolation could be driven by philosophical or ethical considerations, technological advancements that make physical expansion unnecessary, or a preference for inward exploration and self-improvement.

Technological Singularity: Civilizations might develop technologies so advanced that they transcend physical space, making their detection by our current methods impossible. They might exist in a state of post-biological evolution, merging with artificial intelligence or creating virtual realities. Such civilizations could be using energy and resources in ways that are invisible to our current observational capabilities, such as harnessing dark matter or energy, or operating on scales and in dimensions beyond our comprehension.

These possible resolutions illustrate the complexity and multifaceted nature of the Fermi Paradox, reflecting the diverse range of scientific, philosophical, and speculative considerations that it encompasses.

Historical Context and Fermi's Legacy:

Enrico Fermi's question, though simple, struck at the heart of one of the greatest mysteries of the universe. His legacy extends beyond the paradox itself, encompassing contributions to quantum mechanics, nuclear physics, and the development of the first nuclear reactor. The Fermi Paradox, however, remains one of his most enduring legacies, inspiring generations of scientists to ponder the existence of extraterrestrial civilizations and the future of humanity. Fermi's question has not only spurred scientific inquiry but has also influenced popular culture, inspiring countless works of science fiction that explore the possibilities of contact with alien civilizations.

Scientific Investigations and Current Research:

The Fermi Paradox has spurred various scientific investigations and research initiatives. For instance, the SETI Institute continues to scan the skies for radio signals that might indicate the presence of intelligent life. Additionally, projects like the Breakthrough Listen initiative, funded by Yuri Milner, aim to search for technosignatures—indicators of advanced technology—across a wide swath of the universe. These efforts leverage cutting-edge technology and methodologies, such as machine learning algorithms to sift through vast amounts of data for potential signals, and advanced telescopes that can detect minute anomalies in distant star systems.

Recent advancements in astrobiology and planetary science have also contributed to our understanding of the factors that might influence the development and detection of extraterrestrial life. The discovery of extremophiles—organisms that thrive in extreme environments on Earth—has expanded the potential habitats where life might exist beyond our planet. Moreover, the study of exoplanets has revealed a diverse range of planetary environments, some of which may harbor conditions suitable for life. Missions like the James Webb Space Telescope (JWST) and the European Space Agency's PLATO mission are expected to further enhance our ability to detect and study these distant worlds.

Philosophical Implications:

The Fermi Paradox extends beyond scientific inquiry, touching on deep philosophical questions about our place in the universe. It challenges our assumptions about life, intelligence, and the future of civilization. If we are alone, what does that mean for our understanding of life's uniqueness and fragility? If we are not alone, how might contact with another civilization reshape our worldview and future? These questions delve into the realms of ethics, existential risk, and the long-term trajectory of human civilization.

The paradox also invites us to consider the longevity and sustainability of our own civilization. The concept of the Great Filter, in particular, raises questions about the existential risks we face and the steps we must take to ensure our survival and progress. Are we capable of overcoming the challenges that might have led other civilizations to self-destruction, or will we succumb to the same fate? This introspection forces us to confront our priorities and actions on both a personal and societal level, urging us to think deeply about our legacy and our role in the cosmos. Additionally, the potential of self-imposed isolation or the zoo hypothesis makes us reflect on the ethical and philosophical implications of how we treat lesser-developed species and civilizations on Earth. It compels us to consider whether we might one day face a similar choice of isolation versus contact, and what the ethical considerations of such a decision would be.

Conclusion:

The Fermi Paradox remains a profound and unresolved mystery. It challenges us to think critically about the conditions necessary for the emergence of intelligent life and the factors that might limit our ability to detect or communicate with extraterrestrial civilizations. As our technological capabilities advance and our understanding of the universe deepens, we continue to search for answers to Fermi's enduring question: "Where is everybody?" Each hypothesis proposed to resolve the paradox adds a layer of complexity and depth to our understanding, reflecting the multifaceted nature of the search for extraterrestrial life. Whether we find evidence of other civilizations or come to accept our solitude, the journey itself promises to expand our horizons and enrich our understanding of the universe.

THEORY 3:
THE RARE EARTH HYPOTHESIS

COMPLEX LIFE IS A RARE GEM

Introduction to the Rare Earth Hypothesis:

The Rare Earth Hypothesis, formulated by paleontologist Peter Ward and astronomer Donald Brownlee, posits that while microbial life might be common in the universe, complex life forms, like those on Earth, are exceedingly rare. This hypothesis stands in contrast to the more optimistic views of the Drake Equation and suggests that the unique combination of geological, environmental, and astrophysical conditions on Earth is an extraordinarily rare occurrence. Ward and Brownlee introduced this theory in their 2000 book, "Rare Earth: Why Complex Life is Uncommon in the Universe," which has since sparked significant debate and research in the astrobiology community.

The Unique Conditions of Earth:

According to the Rare Earth Hypothesis, the emergence and sustainability of complex life on Earth are the results of a series of fortunate events and conditions. These include:

Stable Climate and Plate Tectonics: Earth's climate has remained relatively stable over geological timescales, in part due to plate tectonics. The movement of tectonic plates helps regulate carbon dioxide levels through the carbon-silicate cycle, preventing runaway greenhouse or icehouse conditions.

Presence of a Large Moon: Earth's large moon plays a crucial role in stabilizing the planet's axial tilt, which moderates seasonal variations and contributes to a stable climate. The moon also influences tides, which may have been important in the development of early life in tidal pools.

Protective Magnetic Field: Earth's magnetic field, generated by its rotating liquid iron core, protects the planet from harmful solar and cosmic radiation. This magnetic shield is vital for maintaining an atmosphere conducive to life.

Location in the Galactic Habitable Zone: Earth is located in a relatively quiet region of the Milky Way, away from the high radiation and gravitational disturbances found in the galaxy's center or spiral arms. This location reduces the risk of catastrophic events such as supernovae, which could disrupt or extinguish life.

Chemical Composition: Earth has a unique chemical composition, with an abundance of heavy elements like carbon, nitrogen, oxygen, and metals essential for complex life. This composition results from the specific history of stellar nucleosynthesis in our region of the galaxy.

The Role of Catastrophic Events:

The Rare Earth Hypothesis also emphasizes the role of catastrophic events in the evolution of life. Mass extinctions, such as the one that wiped out the dinosaurs 65 million years ago, have periodically reset the evolutionary clock, allowing new forms of life to emerge and diversify. These events, while devastating, have created opportunities for the evolution of complexity by eliminating dominant species and enabling adaptive radiations.

For example, the impact hypothesis suggests that the Chicxulub asteroid impact, which caused the Cretaceous-Paleogene extinction event, paved the way for mammals and eventually humans to dominate the planet. Such events are seen as rare but critical in shaping the trajectory of life on Earth.

Scientific Debates and Counterarguments:

While the Rare Earth Hypothesis has garnered significant attention, it remains a subject of intense debate. Critics argue that the hypothesis underestimates the adaptability and resilience of life.

Extremophiles, organisms that thrive in extreme conditions on Earth, suggest that life could exist in a broader range of environments than previously thought. For instance, microorganisms have been found in deep-sea hydrothermal vents, acidic hot springs, and even radioactive waste, indicating that life might thrive in conditions vastly different from those on Earth.

Additionally, recent discoveries of exoplanets in the habitable zones of their parent stars challenge the idea that Earth-like conditions are unique. Missions like Kepler and TESS have identified numerous exoplanets with the potential to support life, suggesting that Earth might not be as rare as the hypothesis proposes. These findings have fueled optimism about the prevalence of habitable worlds in the galaxy.

Implications for the Search for Extraterrestrial Life:

The Rare Earth Hypothesis has significant implications for the search for extraterrestrial life. If complex life is indeed rare, the focus of astrobiology might shift toward finding microbial life, which could be more common. This perspective encourages the exploration of diverse environments within our own solar system, such as the subsurface oceans of Europa and Enceladus, Mars's past habitability, and the methane lakes of Titan.

Furthermore, the hypothesis underscores the importance of planetary protection and the preservation of Earth's unique biosphere. Understanding the factors that make Earth habitable can help us identify and protect similar environments elsewhere, ensuring that we do not inadvertently contaminate them with terrestrial life.

Philosophical Implications:

The Rare Earth Hypothesis also raises profound philosophical questions about the uniqueness of humanity and our place in the cosmos. If complex life is exceedingly rare, Earth and its inhabitants might be the only advanced life forms in the galaxy, imparting a sense of cosmic loneliness and uniqueness. This perspective can deepen our appreciation for the fragility and preciousness of life, emphasizing the need for stewardship of our planet.

Conversely, if future discoveries reveal that complex life is more common than the hypothesis suggests, it could lead to a profound shift in our understanding of the universe and our role within it. The search for extraterrestrial intelligence (SETI) and the potential discovery of advanced civilizations could redefine humanity's self-image and our relationship with the cosmos.

Conclusion:

The Rare Earth Hypothesis offers a compelling and cautionary perspective on the search for extraterrestrial life. By highlighting the unique combination of factors that make Earth habitable, it challenges us to consider the possibility that complex life might be a rare gem in the vast expanse of the universe. Whether or not the hypothesis is ultimately proven correct, it serves as a valuable framework for understanding the conditions necessary for life and guiding future explorations in our quest to answer one of humanity's most profound questions: Are we alone in the universe?

THEORY 4:
THE AESTIVATION HYPOTHESIS

HIBERNATING UNTIL THE UNIVERSE COOLS.

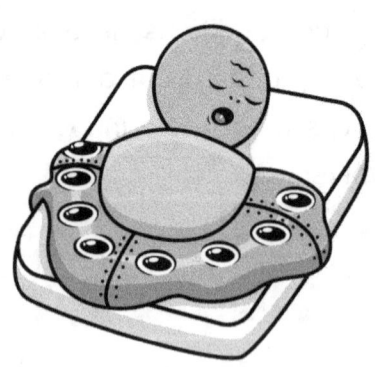

Introduction to the Aestivation Hypothesis

The Aestivation Hypothesis proposes that advanced alien civilizations might be in a state of hibernation or "aestivation," conserving their energy and resources until the universe becomes more conducive to their activities. This idea, which is derived from the concept of aestivation in biology—where animals enter a dormant state during unfavorable environmental conditions—suggests that advanced beings could be waiting for the universe to cool down, making it more energy-efficient for computational processes and other activities. This theory was brought into the scientific spotlight by Anders Sandberg, Stuart Armstrong, and Milan Ćirković in their 2017 paper, which has since sparked considerable interest and debate within the astrobiological community. The hypothesis challenges traditional views of active civilizations by proposing that the most advanced beings might opt for long-term dormancy over continuous activity.

The Concept of Aestivation:

In the natural world, aestivation is a survival strategy used by some animals to avoid extreme heat and desiccation. Similar to hibernation, which is used to escape cold conditions, aestivation allows organisms to enter a state of low metabolic activity until conditions improve. The Aestivation Hypothesis extends this biological concept to advanced extraterrestrial civilizations, suggesting that they might choose to enter a dormant state to preserve energy and resources until the universe reaches a more optimal temperature.

The hypothesis is based on the assumption that the future universe will be significantly cooler than it is today. As the universe expands, the cosmic background temperature will gradually decrease. This cooling will make computational processes, which are assumed to be a primary activity of advanced civilizations, far more energy-efficient. By entering a state of aestivation, these civilizations can avoid the high energy costs associated with current temperatures and wait until the universe cools down to conduct their activities more efficiently. This strategy could allow them to maximize their computational output, which might be crucial for achieving their long-term goals, such as advanced scientific research, complex simulations, or even the creation of new forms of life.

Scientific Basis and Supporting Arguments:

The Aestivation Hypothesis is grounded in several scientific principles and observations:

Thermodynamics and Computational Efficiency: According to the second law of thermodynamics, the efficiency of computational processes increases as the temperature decreases. In a cooler universe, advanced civilizations could perform significantly more computations with the same amount of energy, making it advantageous to wait for these conditions. This principle suggests that by aestivating, civilizations can optimize their use of energy, allowing them to undertake more ambitious and complex projects in the future.

Cosmic Expansion and Cooling: The universe is continuously expanding, leading to a gradual cooling of the cosmic microwave background radiation. Over trillions of years, this cooling will create an environment where energy-intensive processes can be conducted more efficiently. This gradual reduction in temperature can be predicted using current cosmological models, providing a scientific basis for the hypothesis.

Resource Conservation: By entering a dormant state, civilizations can conserve their resources, such as energy and materials, for future use. This strategy could be crucial for long-term survival and technological advancement, allowing civilizations to extend their existence indefinitely. Conserving resources also means that these civilizations could avoid the potential depletion of crucial materials that might be necessary for their future activities.

Implications for the Search for Extraterrestrial Life:

The Aestivation Hypothesis has profound implications for the search for extraterrestrial intelligence (SETI). If advanced civilizations are in a state of aestivation, they would be virtually undetectable by our current methods, which rely on active signals or observable technological artifacts. This dormancy could explain the apparent silence of the cosmos, despite the high probability of advanced civilizations existing somewhere in the universe.

To account for the Aestivation Hypothesis, SETI strategies might need to be adapted to search for indirect signs of dormant civilizations. For example, researchers could look for regions of space with anomalously low entropy or unusually stable conditions that might indicate the presence of a civilization in hibernation. Additionally, advanced technology could be used to detect potential "wake-up" signals—brief bursts of activity that could occur when a civilization temporarily exits its dormant state. These indirect methods would require innovative approaches and the development of new observational technologies capable of detecting subtle anomalies.

Scientific Debates and Counterarguments:

The Aestivation Hypothesis, while intriguing, is not without its critics. Some scientists argue that the hypothesis relies on several speculative assumptions, such as the importance of computational efficiency to advanced civilizations and the feasibility of maintaining a dormant state for extended periods. Additionally, the hypothesis does not account for civilizations that might prioritize exploration, communication, or other activities over energy efficiency.

Another counterargument is that advanced civilizations might find ways to harness energy more effectively in the current universe, rendering aestivation unnecessary. Technologies such as Dyson spheres—massive structures that capture a star's energy—could provide sufficient power for advanced civilizations without the need for dormancy. Furthermore, the potential risks associated with long-term dormancy, such as system failures or cosmic events, might outweigh the benefits. These considerations suggest that while the hypothesis is an interesting one, it may not be the most likely explanation for the silence of the cosmos.

Philosophical Implications:

The Aestivation Hypothesis also raises intriguing philosophical questions about the nature of advanced civilizations and their motivations. If civilizations are willing to enter a state of dormancy for billions of years, what does this say about their values and priorities? Are they more focused on long-term survival and efficiency, or are they driven by goals that are beyond our current understanding?

This hypothesis also challenges our assumptions about progress and activity. It suggests that the most advanced civilizations might not be the ones that are most active or visible, but rather those that are capable of strategic patience and long-term planning.

This perspective can lead us to reevaluate our own priorities and the ways in which we measure technological and societal advancement. The idea of civilizations entering a state of aestivation forces us to consider what it means to be "advanced" and how we define progress, potentially shifting our focus from short-term achievements to long-term sustainability and resilience.

Conclusion:

The Aestivation Hypothesis offers a fascinating and unconventional explanation for the Fermi Paradox. By proposing that advanced civilizations might be in a state of dormancy, it challenges us to think differently about the nature of extraterrestrial life and the factors that influence its detectability. While the hypothesis remains speculative, it provides a valuable framework for exploring new ideas and expanding the scope of our search for extraterrestrial intelligence. As we continue to refine our methods and technologies, the Aestivation Hypothesis will undoubtedly inspire further research and debate, bringing us closer to understanding the true nature of life in the universe. By considering the possibility of aestivating civilizations, we open new avenues for scientific inquiry and philosophical reflection, ultimately enriching our quest to comprehend our place in the cosmos.

THEORY 5:
THE ZOO HYPOTHESIS

OBSERVED BUT UNTOUCHED.

Introduction to the Zoo Hypothesis:

The Zoo Hypothesis proposes that extraterrestrial civilizations are deliberately avoiding contact with humanity to allow for natural evolution and sociocultural development. This theory suggests that advanced beings are observing us in a manner similar to how humans observe animals in a zoo, without interfering or revealing their presence. First articulated by John A. Ball in 1973, the Zoo Hypothesis has become a thought-provoking explanation for the Fermi Paradox, addressing why we have not yet detected signs of intelligent extraterrestrial life despite the high probability of its existence.

The Concept of the Zoo Hypothesis:

The Zoo Hypothesis draws an analogy between how humans study wildlife in controlled environments and how extraterrestrial civilizations might observe humanity. Just as zookeepers avoid direct interaction with animals to allow them to live naturally, advanced civilizations might be refraining from making contact with us to preserve the integrity of our societal and cultural evolution. This hypothesis implies that extraterrestrials possess advanced technology and knowledge that enable them to monitor us without being detected.

This non-interference could be motivated by ethical considerations, similar to the principles behind the Prime Directive in the "Star Trek" universe, which forbids interference with less advanced civilizations. Alternatively, extraterrestrials might consider humanity too primitive or unstable to benefit from contact, preferring to wait until we reach a certain level of technological and ethical maturity.

Scientific Basis and Supporting Arguments:

The Zoo Hypothesis is supported by several key arguments:

Ethical Non-Interference: Advanced civilizations might have developed ethical guidelines that discourage interfering with less advanced civilizations. These guidelines could be based on the understanding that premature contact could disrupt our natural development, potentially causing cultural upheaval or even societal collapse.

Technological Superiority: For the Zoo Hypothesis to hold true, extraterrestrial civilizations would need to possess technology far superior to our own. This technology would enable them to observe us from a distance or remain hidden in plain sight. The rapid advancement of human technology, particularly in stealth and surveillance, suggests that significantly more advanced civilizations could easily remain undetected.

Sociocultural Development: The hypothesis posits that extraterrestrials are waiting for humanity to reach a certain level of development before initiating contact. This threshold could involve advancements in technology, societal stability, or ethical maturity. By allowing us to evolve naturally, extraterrestrials might believe we will be better prepared for eventual contact.

Implications for the Search for Extraterrestrial Life:

The Zoo Hypothesis has significant implications for the search for extraterrestrial intelligence (SETI). If advanced civilizations are intentionally avoiding contact, traditional SETI methods that rely on detecting active signals or technological artifacts might be ineffective. Instead, researchers might need to develop new strategies to detect passive or indirect evidence of extraterrestrial observation.

For example, scientists could look for anomalies in astronomical data that suggest the presence of hidden observers or monitor for patterns that indicate non-random behavior in cosmic phenomena. Additionally, advancements in our own technology might eventually allow us to detect faint signatures or subtle disturbances caused by the presence of advanced extraterrestrial technology.

Scientific Debates and Counterarguments:

While the Zoo Hypothesis offers a compelling explanation for the Fermi Paradox, it also faces several criticisms and counterarguments:

Anthropocentrism: Critics argue that the hypothesis is anthropocentric, projecting human behavior and motivations onto extraterrestrial civilizations. The assumption that advanced beings would act like zookeepers reflects human cultural biases rather than empirical evidence.

Lack of Evidence: The hypothesis is difficult to test because it relies on the absence of evidence as evidence of absence. Without direct contact or observable artifacts, the Zoo Hypothesis remains speculative and unfalsifiable.

Alternative Explanations: Other theories, such as the Great Filter or self-imposed isolation, provide alternative explanations for the lack of contact. These theories suggest that either civilizations do not survive long enough to make contact or choose to remain isolated for reasons unrelated to observing us.

Despite these criticisms, the Zoo Hypothesis continues to be a valuable framework for exploring the potential behavior and motivations of advanced extraterrestrial civilizations.

Philosophical Implications:

The Zoo Hypothesis raises profound philosophical questions about the nature of intelligence and the ethics of contact. If advanced civilizations are observing us without our knowledge, it challenges our understanding of autonomy and free will. The hypothesis also invites us to consider our own behavior towards less advanced species and civilizations on Earth, prompting reflections on our ethical responsibilities.

Furthermore, the idea that we are being observed without interference suggests a form of cosmic stewardship, where advanced beings take a protective role in our development. This perspective can lead to a deeper appreciation for the potential complexity and diversity of life in the universe, as well as a sense of humility regarding our place within it.

Conclusion:

The Zoo Hypothesis offers a thought-provoking and unconventional explanation for the Fermi Paradox. By proposing that advanced civilizations are observing humanity without making contact, it challenges us to rethink our assumptions about extraterrestrial life and the nature of our interactions with the cosmos. While the hypothesis remains speculative, it provides a valuable lens through which to explore the ethical, scientific, and philosophical dimensions of the search for extraterrestrial intelligence. As we continue to develop new technologies and refine our understanding of the universe, the Zoo Hypothesis will undoubtedly inspire further research and debate, contributing to our ongoing quest to uncover the mysteries of the cosmos.

THEORY 6:
PANSPERMIA

SEEDS OF LIFE SCATTERED ACROSS THE STARS.

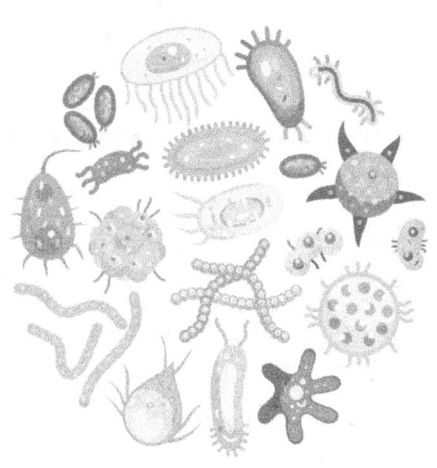

Introduction to Panspermia:

The Panspermia hypothesis suggests that life exists throughout the universe and is distributed by meteoroids, asteroids, comets, planetoids, or, potentially, spacecraft in the form of unintended contamination by microorganisms. This theory posits that life—or at least the precursors of life—originated elsewhere in the cosmos and was transported to Earth, where it eventually evolved into the complex biosphere we see today. First proposed in ancient times and later revived by notable scientists such as Svante Arrhenius and Fred Hoyle, panspermia offers a fascinating perspective on the origins of life and the interconnectedness of the universe.

Historical Background and Development:

The concept of panspermia dates back to ancient Greek philosophers such as Anaxagoras, who speculated that life was distributed throughout the cosmos. In the 20th century, the idea gained scientific traction when Swedish chemist Svante Arrhenius proposed that bacterial spores could travel through space, propelled by radiation pressure. Later, British astronomer Fred Hoyle and his colleague Chandra Wickramasinghe expanded on this idea, suggesting that comets could be carriers of life and that interstellar dust could harbor microorganisms.

Hoyle and Wickramasinghe's work in the 1970s and 1980s revitalized interest in panspermia, proposing that the organic molecules necessary for life were prevalent in space and that comets could deliver these building blocks to planets. Their controversial theory challenged traditional views of life's origins and sparked a wave of research into the possibility of life being more widespread than previously thought.

Mechanisms of Panspermia:

Panspermia can occur through several mechanisms, each with its own implications for the spread of life:

Radiopanspermia: This mechanism, originally proposed by Arrhenius, suggests that microscopic life forms could be propelled through space by radiation pressure from stars. These life forms would need to be highly resistant to the harsh conditions of space, including vacuum, extreme temperatures, and intense radiation.

Lithopanspermia: This hypothesis posits that life can be transported between planets within a star system by rocks ejected from planetary surfaces by impacts. For example, meteorites from Mars have been found on Earth, raising the possibility that life could hitch a ride on such debris. This mechanism is considered one of the most plausible forms of panspermia, given the evidence of meteorite exchange between planets.

Ballistic Panspermia: Similar to lithopanspermia, this mechanism involves the ejection of life-containing rocks from a planet due to volcanic activity or asteroid impacts. These rocks could travel through space and eventually collide with another planet, potentially seeding it with life.

Directed Panspermia: Proposed by Francis Crick and Leslie Orgel, directed panspermia suggests that life was intentionally spread by an advanced extraterrestrial civilization. This idea implies a purposeful seeding of life across the cosmos, raising intriguing questions about the motivations and capabilities of such civilizations.

Supporting Evidence and Research:

Several lines of evidence and research support the plausibility of panspermia:

Survival of Microorganisms in Space: Experiments have shown that some microorganisms, such as tardigrades and certain bacteria, can survive the harsh conditions of space for extended periods. These findings suggest that life forms could endure interplanetary travel if shielded from the most extreme conditions.

Meteorite Studies: Analysis of meteorites has revealed the presence of complex organic molecules, such as amino acids, which are the building blocks of life. These discoveries indicate that the ingredients for life might be widespread in the cosmos and could be delivered to planets via meteoritic material.

Planetary Exchange: The discovery of Martian meteorites on Earth and lunar meteorites on Earth and Mars demonstrates that rocks can travel between planets, providing a potential mechanism for the transfer of life.

Interstellar Organic Molecules: Observations of interstellar space have detected organic molecules, such as polycyclic aromatic hydrocarbons (PAHs), which are considered precursors to life. These findings suggest that the basic components of life might be ubiquitous in the universe.

Implications for the Search for Extraterrestrial Life:

The panspermia hypothesis has significant implications for the search for extraterrestrial life. If life can be transported between planets and star systems, then the search for life should not be limited to finding life that originated independently on other worlds. Instead, scientists should consider the possibility that life on Earth and elsewhere might share a common ancestry.

This perspective encourages a broader approach to astrobiology, focusing on detecting biosignatures and organic molecules in a variety of environments, from the surfaces of comets and asteroids to the icy moons of the outer solar system. Missions like the European Space Agency's Rosetta, which landed on comet 67P/Churyumov-Gerasimenko, and NASA's Perseverance rover on Mars are examples of efforts to uncover the potential for life beyond Earth.

Scientific Debates and Counterarguments:

While panspermia is an intriguing hypothesis, it also faces several scientific challenges and counterarguments:

Survival and Viability: Critics argue that the extreme conditions of space, including intense radiation and the vacuum of space, would likely destroy most microorganisms before they could reach another planet. While some hardy organisms have demonstrated resilience, the vast majority would not survive interstellar travel.

Origin of Life: Panspermia does not explain the origin of life itself; it merely shifts the location of life's emergence. Critics contend that understanding how life originated on Earth is crucial for comprehending the broader question of life's existence in the universe.

Empirical Evidence: While panspermia is supported by some indirect evidence, direct proof of life being transported between planets or star systems remains elusive. More definitive evidence is needed to validate the hypothesis conclusively.

Philosophical Implications:

The panspermia hypothesis also raises profound philosophical questions about the nature of life and our place in the universe. If life is distributed throughout the cosmos, it suggests a deep interconnectedness between all living things, potentially leading to a greater sense of cosmic kinship and unity. This perspective can inspire a broader understanding of life's resilience and adaptability, highlighting the remarkable potential for life to thrive in diverse environments.

Moreover, panspermia challenges the notion of Earth as a unique cradle of life, suggesting instead that our planet might be one of many havens for life in the universe. This idea can lead to a reevaluation of humanity's role and responsibilities in preserving and protecting life, both on Earth and beyond.

Conclusion:

The panspermia hypothesis offers a fascinating and expansive view of the potential distribution of life in the universe. By proposing that life—or its building blocks—can travel between planets and star systems, it challenges traditional notions of life's origins and encourages a broader approach to the search for extraterrestrial life. While the hypothesis remains speculative, it provides a valuable framework for exploring new ideas and expanding our understanding of life's resilience and adaptability. As we continue to investigate the cosmos, the panspermia hypothesis will undoubtedly inspire further research and debate, bringing us closer to unraveling the mysteries of life's origins and distribution.

THEORY 7:
GAIAN BOTTLENECK HYPOTHESIS

SURVIVING THE COSMIC STRUGGLE.

Introduction to the Gaian Bottleneck Hypothesis:

The Gaian Bottleneck Hypothesis, proposed by astrobiologists Aditya Chopra and Charles Lineweaver in 2016, posits that the primary reason we have not found evidence of extraterrestrial civilizations is that life rarely makes it past its initial stages. According to this hypothesis, life on habitable planets frequently emerges but subsequently fails to stabilize and persist long enough to evolve into complex organisms. The term "Gaian Bottleneck" refers to the idea that young planetary environments are often too volatile to sustain life over geological timescales. This hypothesis offers a compelling explanation for the Fermi Paradox by suggesting that while microbial life might be relatively common, the transition to complex life is exceedingly rare.

Concept and Mechanisms:

The Gaian Bottleneck Hypothesis is based on the concept of planetary habitability and the challenges that young planets face in maintaining stable environments conducive to life. Key mechanisms that contribute to the hypothesis include:

Environmental Instability: Young planets are subject to extreme conditions, such as volcanic activity, asteroid impacts, and fluctuating atmospheres. These factors can create hostile environments that make it difficult for life to gain a foothold and stabilize.

Feedback Loops: For a planet to support life over long periods, it must establish feedback mechanisms that regulate the climate and maintain conditions within a narrow, habitable range. Earth has developed such mechanisms, like the carbon-silicate cycle, which helps stabilize temperatures and atmospheric composition. Many planets, however, might fail to develop these self-regulating processes, leading to runaway greenhouse or icehouse conditions that extinguish early life.

Biological Contributions to Habitability: The hypothesis suggests that life itself plays a crucial role in shaping and maintaining habitable conditions. For example, photosynthetic organisms produce oxygen, which in turn supports aerobic life forms. If early life on a planet fails to establish these critical biological contributions, the planet might not sustain habitability.

Supporting Evidence and Research:

Several lines of evidence and research support the Gaian Bottleneck Hypothesis:

Early Earth Conditions: Studies of Earth's history reveal that our planet has undergone numerous catastrophic events and environmental fluctuations. Despite these challenges, life on Earth managed to stabilize and evolve. This resilience might be a rare occurrence, highlighting the difficulties that other planets might face in sustaining life.

Planetary Exploration: Missions to Mars, Venus, and other bodies in the solar system provide insights into the conditions on young planets. Mars, for instance, shows evidence of past liquid water but now has a harsh, dry environment. Venus likely experienced a runaway greenhouse effect, rendering it uninhabitable. These examples illustrate the volatile conditions that young planets might encounter.

Exoplanet Studies: Observations of exoplanets reveal a wide range of planetary environments, many of which are inhospitable. While some exoplanets are located in the habitable zones of their stars, the presence of liquid water and stable climates is not guaranteed. This variability supports the idea that maintaining habitability is a significant challenge.

Introduction to the Gaian Bottleneck Hypothesis:

The Gaian Bottleneck Hypothesis, proposed by astrobiologists Aditya Chopra and Charles Lineweaver in 2016, posits that the primary reason we have not found evidence of extraterrestrial civilizations is that life rarely makes it past its initial stages. According to this hypothesis, life on habitable planets frequently emerges but subsequently fails to stabilize and persist long enough to evolve into complex organisms. The term "Gaian Bottleneck" refers to the idea that young planetary environments are often too volatile to sustain life over geological timescales. This hypothesis offers a compelling explanation for the Fermi Paradox by suggesting that while microbial life might be relatively common, the transition to complex life is exceedingly rare.

Concept and Mechanisms:

The Gaian Bottleneck Hypothesis is based on the concept of planetary habitability and the challenges that young planets face in maintaining stable environments conducive to life. Key mechanisms that contribute to the hypothesis include:

Adaptability of Life: Critics argue that life is highly adaptable and capable of surviving in extreme environments. Extremophiles on Earth, such as those living in hydrothermal vents or acidic hot springs, demonstrate that life can thrive in harsh conditions. This adaptability suggests that life might be more resilient than the hypothesis proposes.

Detection Limitations: The absence of detected extraterrestrial life might be due to the limitations of our current technology rather than the rarity of life. As observational methods improve, we might discover more evidence of life beyond Earth, challenging the Gaian Bottleneck Hypothesis.

Sample Size: The hypothesis is based on a limited sample size, primarily focusing on Earth and a few other bodies in our solar system. With the discovery of more exoplanets and the advancement of planetary science, we might find that the conditions for sustaining life are more common than currently understood.

Philosophical Implications:

The Gaian Bottleneck Hypothesis raises profound philosophical questions about the nature of life and the rarity of complex organisms in the universe. If the hypothesis is correct, it suggests that Earth is an exceptionally rare oasis in a vast, lifeless desert. This perspective can deepen our appreciation for the fragility and uniqueness of life on our planet, emphasizing the importance of environmental stewardship and conservation.

Moreover, the hypothesis challenges the notion of a universe teeming with intelligent civilizations, suggesting instead that the development of complex life is a rare and precious phenomenon. This viewpoint can inspire a sense of cosmic humility, prompting humanity to recognize the unique opportunity we have to explore and understand the universe.

Additionally, the hypothesis prompts reflection on the resilience and perseverance of life. It encourages us to consider the extraordinary series of events and conditions that allowed life on Earth to not only emerge but also thrive over billions of years. This realization can foster a greater respect for the delicate balance required to sustain life and a deeper commitment to protecting it. The Gaian Bottleneck Hypothesis also emphasizes the interconnectedness of life and the environment, highlighting the critical role that living organisms play in maintaining habitable conditions. This interconnectedness can serve as a reminder of our responsibility to preserve the ecosystems and biodiversity that are essential to our planet's health and longevity.

Conclusion:

The Gaian Bottleneck Hypothesis offers a compelling explanation for the Fermi Paradox by proposing that life frequently emerges but rarely stabilizes long enough to evolve into complex organisms. By highlighting the challenges of maintaining habitability on young planets, the hypothesis shifts the focus of astrobiological research toward understanding the conditions that support long-term stability. While the hypothesis remains speculative and subject to ongoing debate, it provides a valuable framework for exploring the resilience and adaptability of life in the universe. As we continue to investigate our own planet's history and explore other worlds, the Gaian Bottleneck Hypothesis will undoubtedly inspire further research and discussion, bringing us closer to unraveling the mysteries of life's persistence and evolution.

THEORY 8:
TECHNOSIGNATURES

FINDING ALIEN TECHNOLOGY, NOT JUST BIOLOGY.

Introduction to Technosignatures:

The search for technosignatures represents an exciting and rapidly evolving frontier in the search for extraterrestrial intelligence (SETI). Technosignatures are indicators of advanced technological activity that could reveal the presence of extraterrestrial civilizations. Unlike biosignatures, which indicate biological processes, technosignatures encompass a wide range of potential evidence, from radio signals and laser emissions to large-scale engineering projects like Dyson spheres. The concept of technosignatures expands the scope of SETI by considering not only the biological but also the technological footprints that intelligent beings might leave behind. This broad approach enhances our chances of detecting signs of extraterrestrial intelligence by incorporating various forms of evidence that go beyond the limitations of conventional biosignature searches.

Types of Technosignatures:

Technosignatures can take many forms, each offering a different method for detecting advanced civilizations:

Radio Signals: One of the earliest and most well-known forms of technosignature is the detection of artificial radio signals. Since the 1960s, projects like the SETI Institute have been scanning the skies for narrow-bandwidth radio signals, which are unlikely to be produced by natural astrophysical processes. These signals could indicate the presence of communication or other technological activities. The vastness of space and the range of frequencies to be scanned make this a challenging but intriguing method of detection, requiring sophisticated equipment and comprehensive analysis techniques.

Laser Emissions: Another potential technosignature is the detection of high-powered laser emissions. Advanced civilizations might use lasers for communication, propulsion, or other purposes. Projects like the Breakthrough Listen initiative have begun incorporating optical SETI to search for these types of signals. The precision and directionality of laser emissions make them an attractive candidate for interstellar communication, suggesting that civilizations capable of such technology might choose this method to minimize signal dispersion and energy loss over vast distances.

Dyson Spheres: Proposed by physicist Freeman Dyson, a Dyson sphere is a hypothetical megastructure that surrounds a star to capture a significant portion of its energy output. Detecting the infrared signature of such a structure could indicate the presence of an advanced civilization. Dyson spheres represent a form of large-scale astroengineering that could significantly alter the energy dynamics of a star system, providing a clear technosignature that is detectable by our current infrared observation capabilities.

Artificial Satellites and Spacecraft: The detection of artificial satellites or spacecraft, particularly those that do not match known human designs or orbits, could serve as a technosignature. This includes unusual objects detected in the vicinity of other stars or unexplained anomalies in the motion of objects within our solar system. Advanced propulsion systems, such as those employing exotic physics or materials, might also leave detectable signatures that differ from natural celestial phenomena.

Pollution and Industrial Byproducts: Advanced civilizations might produce detectable levels of pollutants or industrial byproducts in their atmospheres. For example, the presence of chlorofluorocarbons (CFCs) or other synthetic chemicals in the atmosphere of an exoplanet could indicate technological activity. Such byproducts, while detrimental to the planet's environment, provide a clear signal of industrial processes that would be difficult to attribute to natural causes.

Supporting Evidence and Research:

Several lines of research and evidence support the plausibility of detecting technosignatures:

SETI Projects: The SETI Institute and other organizations have been conducting radio searches for decades, scanning millions of stars for artificial signals. While no definitive technosignatures have been detected, these efforts have refined the techniques and technologies used in the search. Continuous advancements in data processing and signal analysis have enhanced our ability to distinguish potential technosignatures from background noise and natural astrophysical phenomena.

Breakthrough Listen: Funded by Yuri Milner, the Breakthrough Listen initiative represents the most comprehensive search for technosignatures to date. Using some of the world's most powerful radio telescopes, this project scans the nearest million stars and 100 galaxies for potential signals of technological activity. The scope and scale of this initiative, combined with cutting-edge technology, significantly increase the likelihood of detecting technosignatures if they exist within our observational range.

Infrared Searches for Dyson Spheres: Researchers have conducted infrared surveys using telescopes like NASA's WISE (Wide-field Infrared Survey Explorer) to search for the heat signatures of Dyson spheres. While no confirmed detections have been made, these studies provide valuable data on the potential prevalence of such megastructures. By analyzing the infrared spectra of distant stars, scientists can identify anomalies that might suggest the presence of energy-harvesting structures.

Optical SETI: Optical SETI projects, such as those conducted by the LaserSETI initiative, search for brief, intense laser pulses that could indicate communication or other technological activities by extraterrestrial civilizations. These projects utilize advanced optical detectors capable of capturing fleeting signals that might be missed by traditional radio SETI methods.

Implications for the Search for Extraterrestrial Life:

The search for technosignatures has significant implications for SETI and our understanding of the universe:

Broader Search Criteria: By expanding the search beyond biological signatures, scientists can explore a wider range of potential evidence for extraterrestrial life. This holistic approach increases the chances of detecting advanced civilizations. Considering various forms of technosignatures allows researchers to cast a wider net, improving the likelihood of discovering indirect evidence of extraterrestrial intelligence.

Technological Innovation: The search for technosignatures drives technological innovation, leading to the development of advanced sensors, data analysis techniques, and observational strategies. These advancements have broader applications in astronomy and other scientific fields. For instance, improvements in signal detection and processing can enhance our ability to study cosmic phenomena and improve communication technologies.

Interdisciplinary Collaboration: The search for technosignatures requires collaboration across multiple disciplines, including astrophysics, engineering, computer science, and even philosophy. This interdisciplinary approach fosters new ideas and perspectives. The integration of diverse expertise allows for the development of more comprehensive and effective search strategies, as well as deeper philosophical inquiries into the implications of discovering extraterrestrial intelligence.

Scientific Debates and Counterarguments:

While the search for technosignatures is promising, it also faces several challenges and counterarguments:

Signal Ambiguity: Distinguishing between natural and artificial signals can be difficult. Many potential technosignatures could have natural explanations, making it challenging to confirm detections conclusively. The vastness of the universe and the complexity of cosmic phenomena require sophisticated analytical tools to accurately interpret potential technosignatures.

Technological Evolution: Advanced civilizations might use technologies that are currently beyond our understanding or detection capabilities. Our search strategies are based on our own technological development, which might not align with those of extraterrestrial civilizations. This discrepancy highlights the need for adaptable and forward-thinking search methodologies that can accommodate a wide range of technological possibilities.

Fermi Paradox: The lack of detected technosignatures so far contributes to the Fermi Paradox, raising questions about the prevalence and detectability of advanced civilizations. It is possible that civilizations do not use detectable technologies or that they actively avoid detection. This paradox prompts ongoing debate and exploration of alternative explanations for the apparent silence of the cosmos.

Philosophical Implications:

The search for technosignatures also raises profound philosophical questions about the nature of intelligence, technology, and our place in the universe:

Technological Diversity: The search for technosignatures forces us to consider the vast range of possible technologies that advanced civilizations might develop. This diversity challenges our assumptions and broadens our understanding of technological progress. Reflecting on the myriad ways intelligence could manifest technologically encourages a more open-minded and expansive view of the cosmos.

Cosmic Responsibility: If we detect technosignatures, it could lead to ethical questions about our responsibility in the universe. Should we attempt to communicate with these civilizations? What are the potential risks and benefits of such contact? These questions require careful consideration of the ethical and practical implications of interacting with extraterrestrial intelligence, emphasizing the need for responsible and thoughtful decision-making.

Future of Humanity: The search for technosignatures can also prompt reflections on our own technological trajectory and the long-term sustainability of our civilization. By considering the ways in which extraterrestrial civilizations might harness and manage their technologies, we can gain insights into our future development and challenges. This perspective can inspire efforts to ensure the longevity and resilience of human civilization, emphasizing the importance of sustainable and ethical technological advancements.

Conclusion:

The search for technosignatures represents a bold and innovative approach to the search for extraterrestrial intelligence. By expanding the scope of SETI to include technological artifacts and signals, scientists are broadening the horizons of the search and increasing the chances of detecting advanced civilizations. While the search faces significant challenges, the potential discoveries and implications are profound, offering new insights into the nature of intelligence and technology in the universe. As we continue to develop new methods and technologies, the quest for technosignatures will remain a crucial and exciting frontier in our exploration of the cosmos. This endeavor not only advances our understanding of extraterrestrial life but also deepens our appreciation for the intricate interplay between technology, intelligence, and the vast, mysterious universe we inhabit.

THEORY 9:
TRAPPED IN DEEP OCEANS

LIFE HIDDEN UNDER ICY SHELLS.

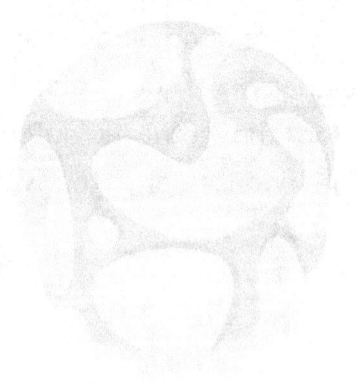

Introduction to the Hypothesis:

The hypothesis that life could be trapped in deep oceans beneath thick ice layers on distant moons and planets offers an intriguing solution to the Fermi Paradox. This theory suggests that life may be more widespread than we realize, thriving in hidden, subsurface oceans shielded from the harsh conditions of space. The potential habitability of these environments is based on discoveries within our own solar system, particularly on moons such as Europa, Enceladus, and Titan. These moons, with their subsurface oceans and energy sources, represent some of the most promising locations for finding extraterrestrial life.

Subsurface Oceans in the Solar System:

Several celestial bodies in our solar system are believed to harbor subsurface oceans beneath their icy crusts. Key examples include:

Europa: One of Jupiter's largest moons, Europa is thought to have a vast ocean beneath its icy surface. The heat generated by tidal flexing due to Europa's interaction with Jupiter keeps this ocean in a liquid state. Europa's ocean is estimated to be in contact with a rocky seafloor, providing the potential for hydrothermal activity, which could supply the energy necessary for life.

Enceladus: Saturn's moon Enceladus has geysers that eject water vapor and ice particles into space, providing direct evidence of a subsurface ocean. Similar to Europa, Enceladus experiences tidal heating, which helps maintain its liquid ocean. The detection of organic molecules in the plumes suggests that Enceladus' ocean could be a habitable environment.

Titan: Saturn's largest moon, Titan, is unique in having surface lakes and seas of liquid methane and ethane. Beneath its thick icy crust, Titan is also believed to have a subsurface ocean of water mixed with ammonia. This environment could provide a stable habitat for life, particularly if it can utilize methane or other hydrocarbons as an energy source.

Ganymede: Another of Jupiter's moons, Ganymede, is thought to contain a layered subsurface ocean sandwiched between layers of ice. Ganymede's magnetic field and geological activity suggest it has a differentiated interior that could support a stable liquid environment.

Mechanisms Supporting Subsurface Life:

Several mechanisms could support life in these deep, subsurface oceans:

Hydrothermal Vents: On Earth, hydrothermal vents on the ocean floor support diverse ecosystems independent of sunlight, relying instead on chemical energy from the planet's interior. Similar hydrothermal systems could exist on icy moons, providing energy and nutrients to sustain microbial life.

Radiolysis: The process of radiolysis involves the dissociation of water molecules by radiation, producing hydrogen and other reactive chemicals. These chemicals can serve as energy sources for microbial life. On icy moons, radiolysis could occur due to the interaction of the moon's ice with cosmic rays or radiation from the planet it orbits.

Chemical Gradients: Subsurface oceans could have chemical gradients created by interactions between the water and the rocky or icy seafloor. These gradients could provide the necessary conditions for chemosynthesis, a process used by some Earth organisms to convert chemical energy into organic matter.

Thermal Energy: Tidal heating generated by gravitational interactions with a parent planet can provide the thermal energy needed to maintain liquid water and drive biochemical processes. This constant supply of heat can create stable, long-term environments suitable for life.

Supporting Evidence and Research:

Research and missions supporting the existence of subsurface oceans and potential life include:

Galileo Mission: The Galileo spacecraft provided the first strong evidence of a subsurface ocean on Europa through its observations of the moon's surface and magnetic field. These findings have been bolstered by subsequent studies and models.

Cassini Mission: The Cassini spacecraft discovered plumes of water vapor erupting from Enceladus and analyzed their composition, revealing organic molecules and providing direct evidence of a subsurface ocean. Cassini's findings have made Enceladus one of the top targets in the search for extraterrestrial life.

Hubble Space Telescope: Observations from the Hubble Space Telescope have detected potential water vapor plumes on Europa, further supporting the presence of a subsurface ocean. These observations have sparked interest in future missions to Europa.

Future Missions: Upcoming missions such as the Europa Clipper and the Jupiter Icy Moons Explorer (JUICE) are designed to explore these icy moons in greater detail, focusing on their subsurface oceans and the potential for life. These missions aim to gather more data on the composition, geology, and habitability of these intriguing environments.

Implications for the Search for Extraterrestrial Life:

The hypothesis that life could be trapped in deep oceans beneath icy shells has profound implications for astrobiology and the search for extraterrestrial life:

Broadening Search Criteria: This hypothesis expands the criteria for habitability beyond the traditional "habitable zone" concept, which focuses on surface liquid water. It suggests that life could exist in a variety of environments, including those that are shielded from harsh surface conditions.

Exploration Priorities: The potential for life in subsurface oceans prioritizes the exploration of icy moons and other bodies with similar characteristics. Missions designed to penetrate the ice and analyze the underlying oceans are crucial for testing this hypothesis.

Detection Techniques: New detection techniques and technologies are needed to explore these hidden environments. This includes developing ice-penetrating instruments, submersible probes, and advanced remote sensing methods to identify biosignatures and chemical anomalies indicative of life.

Scientific Debates and Counterarguments:

While the hypothesis of life in deep subsurface oceans is compelling, it also faces several challenges and counterarguments:

Access and Exploration: Penetrating the thick ice layers covering these oceans poses significant technical challenges. The development of technologies capable of drilling or melting through kilometers of ice is necessary but remains a formidable task.

Energy Sources: The availability of energy sources in these environments is a critical factor. While hydrothermal vents and radiolysis provide potential energy sources, the extent and distribution of these sources are still uncertain.

Extreme Conditions: The conditions in subsurface oceans, such as high pressure, low temperatures, and potential chemical toxicity, could limit the types of life that can survive there. Understanding how life might adapt to these extreme conditions is an ongoing area of research.

Philosophical Implications:

The possibility of life thriving in subsurface oceans beneath icy shells raises profound philosophical questions about the nature of life and the diversity of habitable environments:

Redefining Habitability: This hypothesis challenges our traditional notions of habitable environments, suggesting that life can exist in a broader range of conditions than previously thought. It underscores the adaptability and resilience of life, emphasizing that habitability is not limited to Earth-like conditions.

Interconnectedness of Life: The idea that life could be widespread and connected across different celestial bodies highlights the potential interconnectedness of life in the universe. It prompts us to consider the possibility that life on Earth might share common origins or connections with life elsewhere in the solar system or beyond.

Ethical Considerations: Discovering life in subsurface oceans would raise ethical questions about how we interact with and protect these ecosystems. It would challenge us to consider our responsibilities in preserving extraterrestrial environments and avoiding contamination.

Cosmic Perspective: The search for life in hidden oceans encourages a broader cosmic perspective, reminding us that the universe is vast and full of possibilities. It inspires a sense of wonder and curiosity about the unknown, driving our exploration and understanding of the cosmos.

Human Resilience and Adaptation: Reflecting on how life could adapt to extreme conditions in subsurface oceans can inspire humanity to consider its own resilience and adaptability. Understanding that life can thrive in such harsh environments can lead to a greater appreciation for the potential of life to survive and flourish in diverse conditions, including the extreme environments we might face on Earth and in space exploration.

Conclusion:

The hypothesis that life could be trapped in deep oceans beneath thick ice layers on distant moons and planets offers a fascinating solution to the Fermi Paradox and broadens our understanding of habitable environments. By exploring subsurface oceans on moons like Europa, Enceladus, Titan, and Ganymede, scientists are expanding the search for extraterrestrial life and uncovering new possibilities for where life might thrive. While significant challenges remain in accessing and studying these hidden environments, ongoing and future missions hold the promise of exciting discoveries. As we continue to explore the cosmos, the hypothesis of life in subsurface oceans will undoubtedly inspire further research and debate, bringing us closer to answering one of humanity's most profound questions: Are we alone in the universe?

THEORY 10:
HYDROGEN-BREATHING LIFE FORMS

LIVING WITHOUT OXYGEN.

Introduction to Hydrogen-Breathing Life Forms:

The hypothesis of hydrogen-breathing life forms suggests that life could exist in environments where hydrogen is the primary electron donor, rather than oxygen. This theory expands our understanding of the possible biochemistries that could support life, especially in environments vastly different from Earth. It posits that extraterrestrial life might thrive in hydrogen-rich atmospheres, potentially broadening the search for habitable worlds beyond those with oxygenated atmospheres.

Basis of the Hypothesis:

The hypothesis is grounded in several scientific observations and theoretical considerations:

Diversity of Life on Earth: Life on Earth exhibits remarkable biochemical diversity, with some organisms capable of metabolizing hydrogen. For instance, certain extremophiles—organisms that thrive in extreme environments—use hydrogen as a primary energy source. This adaptability suggests that life elsewhere might also utilize hydrogen in ways that are not common on Earth.

Hydrogen as an Energy Source: Hydrogen is the most abundant element in the universe, making it a plausible energy source for extraterrestrial life. Hydrogen metabolism involves the use of hydrogen as an electron donor in various biochemical processes, a method utilized by some anaerobic bacteria and archaea on Earth.

Astrobiological Potential: The discovery of hydrogen-rich environments on celestial bodies such as Saturn's moon Enceladus, Jupiter's moon Europa, and even exoplanets, supports the potential for hydrogen-based life. These environments offer the necessary conditions for hydrogen metabolism, including liquid water and the presence of necessary chemical nutrients.

Mechanisms of Hydrogen Metabolism:

Hydrogen-breathing life forms could utilize several biochemical mechanisms for energy production:

Hydrogen Oxidation: This process involves the oxidation of hydrogen gas (H_2) to produce energy. On Earth, certain microorganisms such as hydrogenotrophic methanogens perform hydrogen oxidation, combining hydrogen with carbon dioxide to produce methane and energy.

Hydrogen-Driven Chemosynthesis: In environments where sunlight is absent, such as deep-sea hydrothermal vents or subsurface oceans on icy moons, chemosynthetic organisms can use hydrogen as an energy source. These organisms rely on chemical reactions, rather than photosynthesis, to produce organic compounds and sustain life.

Alternative Electron Acceptors: While oxygen is a common electron acceptor in many biological processes, hydrogen-breathing life forms might use other elements such as sulfur, nitrogen, or carbon dioxide. The flexibility in electron acceptors allows for a wider range of possible life-supporting environments.

Supporting Evidence and Research:

Several lines of evidence and research support the hypothesis of hydrogen-breathing life forms:

Extremophiles on Earth: The discovery of extremophiles that metabolize hydrogen in extreme environments on Earth provides a proof of concept for hydrogen-based life. These organisms thrive in environments such as hydrothermal vents, acidic hot springs, and deep subsurface ecosystems, demonstrating the viability of hydrogen metabolism.

Astrobiological Missions: Missions such as Cassini and Galileo have detected hydrogen in the plumes of Enceladus and the suspected subsurface oceans of Europa. These findings suggest that hydrogen is available in environments that could support life.

Laboratory Studies: Experimental studies have shown that certain microorganisms can thrive in hydrogen-rich environments. Research on anaerobic bacteria and archaea has expanded our understanding of the biochemical pathways that support hydrogen metabolism.

Implications for the Search for Extraterrestrial Life:

The hypothesis of hydrogen-breathing life forms has significant implications for the search for extraterrestrial life:

Expanding Habitability Criteria: This hypothesis broadens the criteria for habitability beyond Earth-like conditions. It suggests that hydrogen-rich environments, even those lacking oxygen, could support life, thus expanding the range of celestial bodies considered potentially habitable.

Targeting Hydrogen-Rich Worlds: Future astrobiological missions might prioritize the exploration of hydrogen-rich environments. Instruments designed to detect hydrogen and associated biosignatures will be crucial for identifying potential habitats for hydrogen-breathing life forms.

Rethinking Biosignatures: Traditional biosignatures focus on detecting oxygen and carbon-based life. The hydrogen-breathing hypothesis encourages the development of new biosignature detection techniques tailored to hydrogen-based metabolisms.

Scientific Debates and Counterarguments:

While the hypothesis of hydrogen-breathing life forms is compelling, it also faces several challenges and counterarguments:

Metabolic Efficiency: Critics argue that hydrogen metabolism might be less efficient than oxygen-based respiration. The energy yield from hydrogen oxidation is generally lower, which could limit the complexity and diversity of hydrogen-breathing life forms.

Environmental Constraints: The availability of hydrogen and other necessary nutrients might be limited in many environments, posing a constraint on the potential for hydrogen-based life. Understanding the distribution and abundance of hydrogen is crucial for evaluating this hypothesis.

Detection Challenges: Detecting hydrogen-breathing life forms poses significant technical challenges. Current astrobiological instruments are primarily designed to detect oxygen and carbon-based biosignatures. Developing new technologies for detecting hydrogen-based life will require substantial research and innovation.

Philosophical Implications:

The possibility of hydrogen-breathing life forms raises profound philosophical questions about the nature of life and our search for extraterrestrial intelligence:

Redefining Life: This hypothesis challenges our traditional definitions of life and the conditions necessary for its existence. It prompts us to consider a broader spectrum of biochemical possibilities, expanding our understanding of what it means to be alive.

Interconnectedness of Life: The potential for hydrogen-breathing life emphasizes the adaptability and resilience of life, suggesting that life could thrive in a variety of environments. This interconnectedness highlights the universality of life's fundamental principles, regardless of the specific biochemistry.

Ethical Considerations: Discovering hydrogen-breathing life forms would raise ethical questions about our responsibilities in preserving and protecting these ecosystems. It challenges us to consider the implications of our exploration and potential impact on extraterrestrial life.

Cosmic Perspective: The search for hydrogen-breathing life forms encourages a broader cosmic perspective, reminding us of the vastness and diversity of the universe. It inspires a sense of wonder and curiosity about the unknown, driving our exploration and understanding of the cosmos.

Human Adaptation: Reflecting on the possibility of life thriving in hydrogen-rich environments can inspire humanity to consider its own adaptability. Understanding that life can survive in such diverse conditions emphasizes the potential for human exploration and settlement in similarly extreme environments, both on Earth and beyond.

Conclusion:

The hypothesis of hydrogen-breathing life forms offers a fascinating and expansive view of the potential diversity of life in the universe. By considering the possibility that life could thrive in hydrogen-rich environments, scientists are expanding the search for extraterrestrial life and uncovering new possibilities for habitability. While significant challenges remain in detecting and studying hydrogen-based life, ongoing and future research hold the promise of exciting discoveries. As we continue to explore the cosmos, the hypothesis of hydrogen-breathing life forms will undoubtedly inspire further research and debate, bringing us closer to answering one of humanity's most profound questions: Are we alone in the universe?

THEORY II:
BLACK HOLE HABITATS

THRIVING NEAR COSMIC GIANTS.

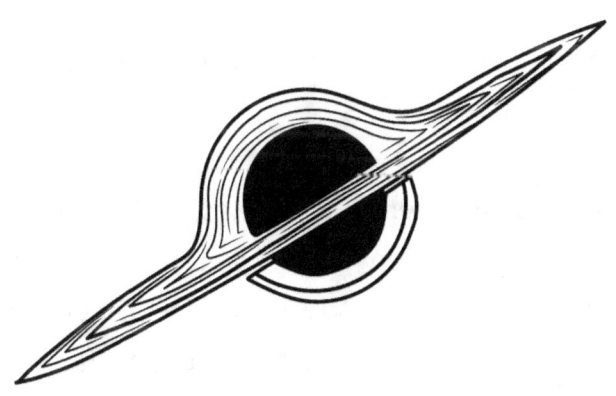

Introduction to Black Hole Habitats:

The Black Hole Habitats theory proposes that advanced extraterrestrial civilizations might exist near black holes, utilizing the immense gravitational and energy resources they provide. This theory suggests that the regions around black holes, particularly the event horizon and accretion disk, could offer unique opportunities for energy extraction and scientific experimentation, potentially supporting advanced forms of life or even civilizations.

Basis of the Hypothesis:

The hypothesis is based on several key concepts and observations:

Energy Abundance: Black holes, especially those with active accretion disks, generate tremendous amounts of energy. Advanced civilizations could harness this energy using technologies far beyond our current capabilities. The idea of extracting energy from black holes is often associated with the theoretical concept of a Penrose process, where energy is extracted from the rotational energy of a black hole.

Time Dilation: Near a black hole, time dilation effects due to intense gravitational fields could be significant. For advanced civilizations, this could provide advantages in terms of conducting long-term scientific experiments, preserving information, or experiencing different rates of time compared to other parts of the universe.

Stable Orbits: It is theoretically possible for planets or artificial habitats to exist in stable orbits around a black hole. These orbits could provide relatively stable environments where life or civilizations might thrive, protected by advanced technology.

Supporting Arguments and Research:

Several lines of research and theoretical work support the plausibility of black hole habitats:

Penrose Process: The Penrose process describes how energy can be extracted from a rotating black hole. This theoretical framework provides a basis for understanding how advanced civilizations might exploit the energy resources of black holes.

Accretion Disks: The intense radiation and energy from accretion disks around black holes could be harnessed by advanced technologies. This environment, while hostile to life as we know it, could be manipulated or shielded by sufficiently advanced civilizations.

Astrophysical Observations: Observations of quasars and other active galactic nuclei, which are powered by supermassive black holes, demonstrate the enormous energy outputs that black holes can generate. These observations suggest that black holes are not only destructive forces but also potential sources of immense energy.

Implications for the Search for Extraterrestrial Life:

The hypothesis of black hole habitats has several implications for the search for extraterrestrial intelligence:

Broadening Search Areas: Traditional searches for extraterrestrial life often focus on Earth-like planets within habitable zones of stars. The black hole habitats theory suggests that we should also consider regions near black holes, despite their extreme conditions.

Technological Signatures: Advanced civilizations near black holes might leave detectable technological signatures, such as unusual radiation patterns or artificial structures. Searching for these signatures could provide new avenues for SETI.

Astrobiological Exploration: Understanding how life or civilizations could exist near black holes could expand our knowledge of the possible diversity of life and technological advancement in the universe.

Scientific Debates and Counterarguments:

While the idea of black hole habitats is intriguing, it also faces several challenges and counterarguments:

Hostile Environments: The regions near black holes are extremely hostile, with intense radiation, tidal forces, and gravitational gradients. These conditions would require advanced technologies to mitigate and might limit the feasibility of life or habitats in these areas.

Technological Limitations: Our current understanding of physics and technology is insufficient to fully grasp how energy extraction or stable habitats near black holes could be achieved. The theoretical nature of the hypothesis makes it difficult to test with existing technology.

Alternative Explanations: Some argue that there are more plausible locations for advanced civilizations to thrive, such as around stable stars or within large, resource-rich galaxies. The extreme nature of black hole environments might make them less favorable compared to other, more stable environments.

Conclusion:

The Black Hole Habitats theory offers a fascinating and unconventional perspective on the potential locations of advanced extraterrestrial civilizations. By considering the unique energy and scientific opportunities provided by black holes, this hypothesis challenges traditional views of habitable environments and expands the scope of our search for extraterrestrial life. While significant challenges remain in understanding and exploring these extreme regions, the idea of civilizations thriving near cosmic giants continues to inspire scientific and philosophical inquiry.

THEORY 12:
THE TRANSIENT HYPOTHESIS

BRIEFLY EXISTING CIVILIZATIONS.

Introduction to the Transient Hypothesis:

The Transient Hypothesis posits that civilizations throughout the universe may emerge, develop, and eventually self-destruct or decline within relatively short timescales, making it difficult for them to overlap and communicate with each other. This hypothesis offers a potential explanation for the Fermi Paradox by suggesting that the window of time during which civilizations are detectable is exceedingly brief.

Basis of the Hypothesis

The hypothesis is grounded in several key observations and theoretical considerations:

Technological and Societal Instability: Civilizations may face numerous challenges that threaten their longevity, including environmental degradation, resource depletion, nuclear war, or uncontrolled technological advancements. These factors could lead to the collapse or transformation of civilizations before they have the chance to establish long-term communication with others.

Rapid Technological Evolution: The rapid pace of technological change could result in civilizations transitioning quickly from detectable forms of communication (like radio waves) to more advanced or undetectable methods. As civilizations evolve, they might abandon detectable technologies, reducing the likelihood of overlapping with other civilizations using similar technologies.

Existential Risks: Advanced civilizations may face existential risks, such as artificial intelligence, biotechnology, or cosmic events (like supernovae or gamma-ray bursts). These risks could lead to the extinction or drastic reduction of intelligent life, shortening the period during which they can communicate with others.

Supporting Arguments and Research:

Several lines of reasoning and research support the Transient Hypothesis:

Historical Precedents: Human history shows multiple examples of civilizations rising and falling within a few centuries. Ancient empires such as Rome, the Maya, and the Gupta Empire flourished and declined, often due to a combination of internal and external pressures. These historical precedents suggest that the lifespan of civilizations may inherently be limited.

Technological Trends: Observations of technological trends indicate that advancements often lead to both progress and new risks. The development of nuclear weapons, for example, created unprecedented destructive potential alongside technological achievements. Similarly, emerging technologies like artificial intelligence pose significant risks that could lead to rapid societal changes or even collapse.

Astrobiological Considerations: Studies of astrobiological factors, such as the habitability of exoplanets and the potential for cosmic catastrophes, highlight the fragility of life in the universe. Factors like asteroid impacts, supernovae, and climate instability can drastically affect the development and sustainability of civilizations.

Implications for the Search for Extraterrestrial Life:

The Transient Hypothesis has significant implications for the search for extraterrestrial intelligence (SETI):

Temporal Constraints: This hypothesis emphasizes the importance of considering the temporal aspects of civilization detection. SETI efforts might need to account for the possibility that civilizations are detectable only during brief windows of time, necessitating continuous and widespread monitoring.

Technological Evolution: Understanding the technological evolution of civilizations can help refine search strategies. Researchers should consider both emerging and declining technologies, focusing on a broad spectrum of potential signals rather than relying solely on traditional methods like radio astronomy.

Interdisciplinary Approaches: Addressing the Transient Hypothesis requires interdisciplinary approaches that integrate insights from history, sociology, and technology studies. By examining the factors that influence the rise and fall of civilizations, researchers can develop more effective strategies for detecting and understanding extraterrestrial life.

Scientific Debates and Counterarguments:

While the Transient Hypothesis is compelling, it also faces several challenges and counterarguments:

Detection Limitations: Critics argue that the lack of detected extraterrestrial signals could be due to our limited observational capabilities rather than the transient nature of civilizations. Technological advancements in detection methods might reveal previously undetectable signals.

Survivability and Adaptation: Some researchers believe that advanced civilizations could develop strategies to mitigate existential risks and ensure long-term survival. Concepts like space colonization, artificial habitats, and robust planetary defenses could extend the lifespan of civilizations.

Sample Size: The hypothesis is based on a limited sample size, primarily focusing on human history and technological trends. The experiences of other civilizations, if fundamentally different from our own, could lead to different outcomes and longer periods of detectability.

Philosophical Implications:

The Transient Hypothesis raises profound philosophical questions about the nature of civilization and the factors that influence its longevity:

Impermanence of Civilizations: This hypothesis challenges the notion of permanence, suggesting that all civilizations are inherently transient. It prompts reflections on the cyclical nature of existence and the factors that contribute to the rise and fall of societies.

Ethical Considerations: Understanding the transient nature of civilizations raises ethical questions about our responsibilities to future generations and the broader cosmic community. It emphasizes the importance of sustainable development and the need to mitigate existential risks.

Interconnectedness and Cooperation: The hypothesis underscores the interconnectedness of civilizations and the potential benefits of cooperation. It encourages a global perspective, highlighting the need for collaborative efforts to address shared challenges and ensure the long-term survival of intelligent life.

Existential Reflection: Reflecting on the Transient Hypothesis can lead to a deeper appreciation for the fragility and preciousness of civilization. It encourages a sense of humility and responsibility, prompting us to consider how we can extend the lifespan of our own civilization and contribute to the broader cosmic community.

Cultural Legacy: The potential impermanence of civilizations raises questions about the legacy we leave behind. It prompts us to consider the cultural, scientific, and ethical contributions we make to the universe and how they might influence future civilizations, whether on Earth or elsewhere.

Conclusion:

The Transient Hypothesis offers a compelling explanation for the Fermi Paradox by suggesting that civilizations may be brief and difficult to detect due to their transient nature. By emphasizing the temporal constraints and existential risks faced by advanced civilizations, this hypothesis challenges traditional views of civilization longevity and expands the scope of our search for extraterrestrial life. While significant challenges remain in understanding and addressing these temporal dynamics, ongoing research and interdisciplinary approaches hold the promise of uncovering new insights. As we continue to explore the cosmos, the Transient Hypothesis will undoubtedly inspire further research, debate, and philosophical reflection, driving us to consider new possibilities about the nature and longevity of civilizations in the universe.

Conclusion

The journey through these twelve theories about extraterrestrial life has expanded our understanding of the cosmos and our place within it. From estimating the number of civilizations with the Drake Equation to contemplating the potential for brief, transient civilizations, each theory offers a unique perspective on the Fermi Paradox and the search for extraterrestrial intelligence.

Our exploration began with the **Drake Equation**, a mathematical framework for estimating the number of detectable civilizations in our galaxy. We then delved into the **Fermi Paradox**, which highlights the apparent contradiction between high estimates of extraterrestrial civilizations and the lack of evidence for or contact with such civilizations. The **Rare Earth Hypothesis** suggested that complex life might be exceptionally rare due to a unique combination of geological and environmental conditions required for its development.

We also explored the **Aestivation Hypothesis**, proposing that advanced civilizations might be in a state of hibernation, waiting for more favorable cosmic conditions. The **Zoo Hypothesis** presented the idea that extraterrestrial civilizations might deliberately avoid contact with us to allow for our natural development. **Panspermia** expanded our view of life's origins, suggesting that life might be widespread in the universe, distributed by meteoroids, comets, and other celestial bodies.

The **Gaian Bottleneck Hypothesis** posited that while microbial life might be common, complex life is rare because many young planets fail to maintain stable environments. The search for **Technosignatures** broadened our methods for detecting advanced civilizations, looking for technological artifacts and signals beyond biological indicators. The hypothesis of life **Trapped in Deep Oceans** on icy moons like Europa and Enceladus opened new frontiers in astrobiology, suggesting that subsurface oceans could harbor life.

The idea of **Hydrogen-Breathing Life Forms** challenged our Earth-centric view of life, proposing that life could thrive in hydrogen-rich environments. The concept of **Black Hole Habitats** suggested that advanced civilizations might utilize the immense energy resources of black holes. Finally, the **Transient Hypothesis** highlighted the possibility that civilizations might rise and fall within short timescales, making it challenging to detect them.

As we conclude this exploration, it is clear that the universe is full of possibilities, and our search for extraterrestrial life continues to push the boundaries of science and imagination. Each theory not only brings us closer to answering the profound question of whether we are alone in the universe but also enriches our understanding of life's resilience and adaptability.

For readers interested in further expanding their knowledge about space exploration and the future of humanity, we recommend our other books:

- **"20 Habitable Planets: Discovering New Homes for Humanity"** explores the most promising exoplanets that could potentially support human life, delving into the scientific discoveries and technological advancements that bring us closer to finding new homes beyond Earth.

- **"Beyond Earth: The Rise of Private Space Travel"** examines the burgeoning industry of private space travel, highlighting the key players, technological innovations, and future prospects that are transforming space exploration from a governmental endeavor to a commercial enterprise.

- **"Mars Colonization: The Red Planet New Era"** provides a comprehensive look at the plans and missions aimed at establishing human presence on Mars. It covers the challenges, scientific breakthroughs, and visionary projects that are paving the way for a new era of exploration on the Red Planet.

As we continue our quest to understand the cosmos, these works will guide you through the exciting developments and possibilities that lie ahead. Together, we stand on the brink of discovering new worlds and unraveling the mysteries of the universe.

Thank you for joining us on this journey through the fascinating theories about extraterrestrial life. May your curiosity and passion for exploration continue to drive you towards new discoveries and a deeper appreciation of the vast, mysterious universe we call home.

Glossary of Terms

Accretion Disk: A structure formed by diffused material in orbital motion around a central body, such as a star or black hole. The material spirals inward, often emitting energy as it heats up.

Astrobiology: The study of the origin, evolution, distribution, and future of life in the universe. This interdisciplinary field combines aspects of astronomy, biology, geology, and chemistry.

Biosignature: Any substance—such as an element, isotope, molecule, or phenomenon—that provides scientific evidence of past or present life. Examples include atmospheric gases like oxygen or methane that could indicate biological activity.

Drake Equation: A formula developed by Dr. Frank Drake in 1961 to estimate the number of active, communicative extraterrestrial civilizations in the Milky Way galaxy. The equation considers factors like the rate of star formation, the fraction of stars with planetary systems, and the likelihood of life developing.

Dyson Sphere: A hypothetical megastructure that encompasses a star to capture a significant portion of its power output. This concept was proposed by physicist Freeman Dyson as a way for advanced civilizations to harness the energy of their stars.

Extremophiles: Organisms that thrive in extreme environmental conditions, such as high temperatures, acidity, salinity, or pressure. Examples include bacteria living in hydrothermal vents or acidic hot springs.

Fermi Paradox: The apparent contradiction between the high probability of extraterrestrial civilizations existing in the universe and the lack of evidence or contact with such civilizations. Named after physicist Enrico Fermi, who famously asked, "Where is everybody?"

Gaian Bottleneck Hypothesis: A theory suggesting that while microbial life might frequently arise on habitable planets, complex life is rare because many young planets fail to maintain stable environments long enough for life to evolve and thrive.

Habitable Zone: Also known as the "Goldilocks Zone," this is the region around a star where conditions are suitable for liquid water to exist on a planet's surface, which is considered essential for life as we know it.

Hydrothermal Vents: Openings in the sea floor that emit hot, mineral-rich water. These environments support diverse ecosystems independent of sunlight, relying instead on chemical energy from the Earth's interior.

Panspermia: The hypothesis that life exists throughout the universe and is distributed by space dust, meteoroids, comets, and planetoids. It suggests that life on Earth may have originated from microorganisms or chemical precursors of life present in outer space.

Radiolysis: The dissociation of molecules by radiation, which can produce reactive chemicals that may serve as energy sources for life. This process could support microbial life in environments like subsurface oceans on icy moons.

SETI (Search for Extraterrestrial Intelligence): A collective term for scientific efforts to detect signs of intelligent extraterrestrial life, typically by monitoring electromagnetic radiation for signals that could indicate technologically advanced civilizations.

Technosignature: Any evidence of technology that suggests the presence of an advanced civilization. This can include radio signals, laser emissions, artificial satellites, and megastructures like Dyson spheres.

Tidal Heating: A process where gravitational forces from a planet and its moon (or another nearby celestial body) cause internal friction and heat within the moon, potentially maintaining subsurface oceans in a liquid state.

Transient Hypothesis: The theory that civilizations may emerge, develop, and eventually self-destruct or decline within relatively short timescales, making it difficult for them to overlap and communicate with each other.

Zoo Hypothesis: The idea that advanced extraterrestrial civilizations are aware of humanity but avoid contact to allow for natural evolution and sociocultural development, similar to zookeepers observing animals in a zoo.

This glossary provides concise definitions of key terms and concepts discussed in the book, helping readers quickly reference and understand the scientific ideas presented.

References

Bostrom, Nick. "**Are You Living in a Computer Simulation?**" Philosophical Quarterly, 2003. This paper discusses the Simulation Hypothesis, proposing that it is more likely than not that we are living in a computer simulation created by an advanced civilization.

Chopra, Aditya, and Charles H. Lineweaver. "**The Case for a Gaian Bottleneck: The Biology of Habitability.**" Astrobiology, 2016. This paper presents the Gaian Bottleneck Hypothesis, suggesting that while microbial life may be common, complex life is rare due to the instability of young planets.

Dyson, Freeman. "**Search for Artificial Stellar Sources of Infrared Radiation.**" Science, 1960. In this paper, Dyson proposes the concept of Dyson spheres, megastructures built by advanced civilizations to capture a star's energy.

Frank, Adam, and Woodruff Sullivan. "**A New Empirical Constraint on the Prevalence of Technological Species in the Universe.**" Astrobiology, 2016. This paper discusses the statistical probabilities related to the Drake Equation and the search for extraterrestrial intelligence (SETI).

Sandberg, Anders, Stuart Armstrong, and Milan Ćirković. "**That is not dead which can eternal lie: the aestivation hypothesis for resolving Fermi's paradox.**" Journal of the British Interplanetary Society, 2017. This paper explores the Aestivation Hypothesis, suggesting advanced civilizations may be in a state of hibernation, conserving energy until the universe cools.

Wickramasinghe, Chandra, and Fred Hoyle. **Evolution from Space. J.M. Dent & Sons, 1981.** This book discusses the Panspermia hypothesis, proposing that life on Earth may have originated from microorganisms or chemical precursors of life present in outer space.

Webb, Stephen. **If the Universe Is Teeming with Aliens... Where Is Everybody? Fifty Solutions to the Fermi Paradox and the Problem of Extraterrestrial Life.** Copernicus Books, 2002. This book provides a comprehensive overview of various hypotheses, including the Zoo Hypothesis and the Fermi Paradox, related to the search for extraterrestrial intelligence.

Europa Clipper Mission. NASA, europa.nasa.gov This upcoming mission aims to explore Jupiter's moon Europa, investigating its subsurface ocean and potential for life.

Breakthrough Listen Initiative. breakthroughinitiatives.org/initiative This initiative is the most comprehensive search for technosignatures, using advanced telescopes to scan the nearest million stars and 100 galaxies for signs of intelligent life.

These references provide the foundational literature and contemporary research supporting the various theories discussed in the book, offering readers avenues for further exploration into the fascinating field of astrobiology and the search for extraterrestrial life.

www.ingramcontent.com/pod-product-compliance
Lightning Source LLC
Chambersburg PA
CBHW071518220526
45472CB00003B/1061